written by
Emer Martin
and
Suzana Tulac, Ph.D.

Why is the
Moon
Following Me?

illustrated by
Magdalena Zuljevic

RAWMEASH

This book is dedicated to all of our kids, Nikola, Maya, Jasmine, Jade, Athena, Astra, Ada, and all the other curious children who love looking at the stars.

ACKNOWLEDGEMENTS:

We would like to thank a few people who helped make this book possible: Alka Raghuram, Anna Van Lenton, Maria Behan, Jasmine Martin-Partovi, and Lisie Sabbag for invaluable help editing; Holly Brady for giving us good advice and recommendations; Diana Russell for her great design. Finally, to our long suffering partners Zeljko, Afshin, and Sanjin for believing in this project, and understanding that we weren't just meeting to drink tea and chat.

Why Is The Moon Following Me?
Poems copyright © 2014 by Emer Martin.
Early Astronomers copyright © 2014 by Suzana Tulac.
Illustrations copyright © 2014 by Magdalena Zuljevic.

Author Photos by Suzana Tulac suzanatulac.wix.com/stulacphotography
Cover design Magdalena Zuljevic

ISBN 978-0-9913547-1-9
Printed in U.S.A.
First Edition 2014, Rawmeash Books Edition
www.rawmeash.com

EARLY ASTRONOMERS

Aristotle
(384–322 BC)

A Greek philosopher, one of the greatest thinkers in all of ancient civilizations, whose studies laid the foundation for scientific thought.

Aristarchus
(~310–230 BC)

A Greek astronomer and mathematician, the first to present the heliocentric model with the sun at the center of the universe and the earth revolving around it.

Ptolemy
(~87–170 AD)

A Greek mathematician, astronomer and geographer from Alexandria, who proposed the first general theory of cosmology.

Nicolaus Copernicus
(1473–1543)

A Polish renaissance mathematician and astronomer who proposed a heliocentric model of the solar system in which the planets orbit around the sun.

Galileo Galilei
(1564–1642)

An Italian astronomer, physicist and mathematician who played a major role in the beginning of the movement known as the 'Scientific Revolution'.

Tycho Brahe
(1546–1601)

A Danish astronomer and alchemist who made a profound contribution in astronomical and planetary observations.

Johannes Kepler
(1571–1630)

A German mathematician and astronomer who explained planetary motion.

Since time began, all young earthlings
Gazed far, far up at the great skies circling.
They loved to watch the magical night sky;
And so the curious child kept asking WHY?
Why is the moon following me?
Is all there is, all I see?
What are the stars?

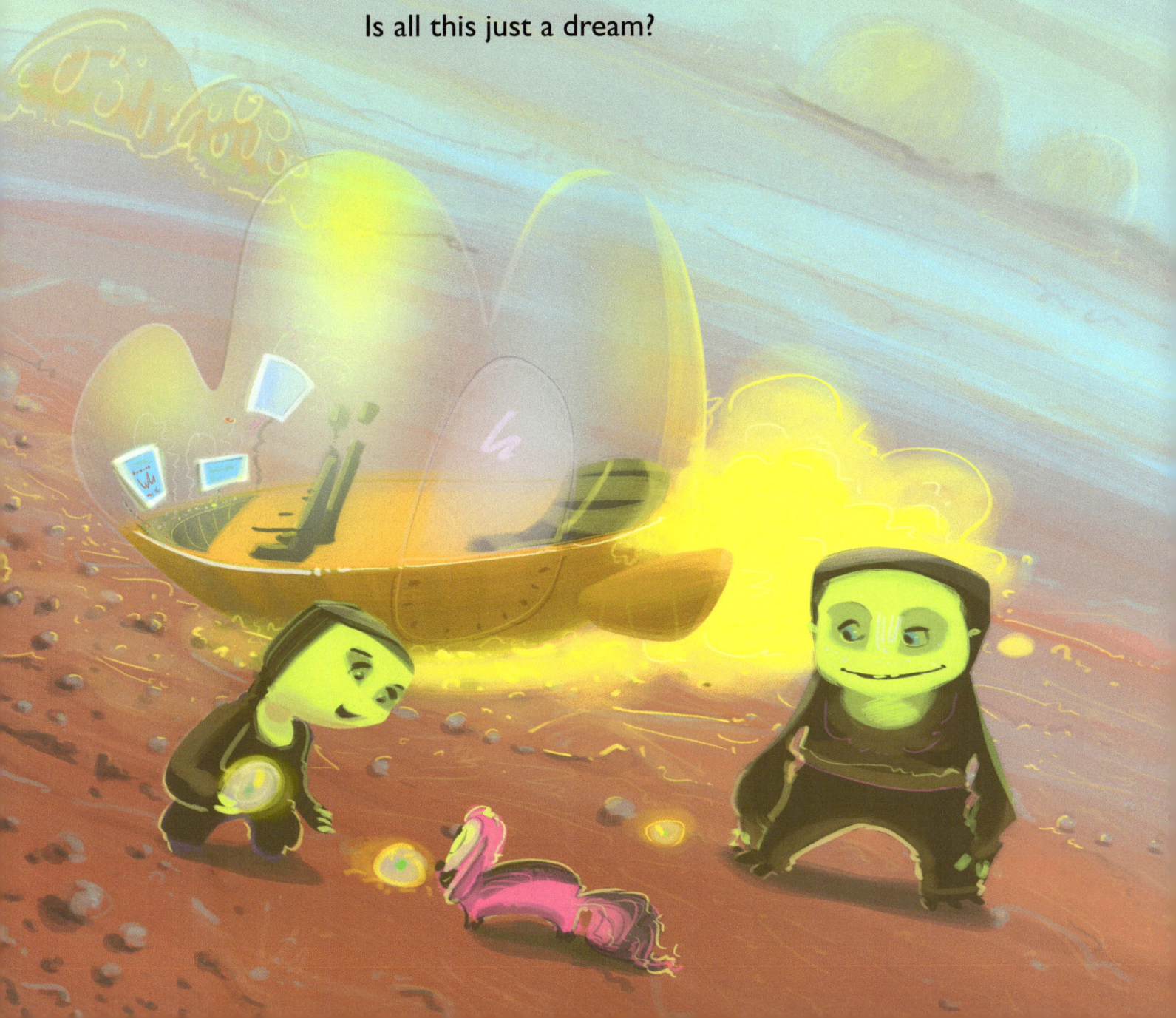

Is there life on Mars?

Where does the sun go to sleep?

How does the blue sky turn black and deep?

Can we go there?

Is there anywhere?

Would eating the Milky Way be fun?

Can I swim in the sun?

If aliens visit me would I scream?

Is all this just a dream?

Since time began, the mothers said,
"Stop that nonsense and go to bed.
Lying there with your brain all fired,
In the morning you'll be tired."
But once the world was drained of light
Imagination could take flight.
Around ancient fires clever children curled
While their young minds travelled out of this world.

Now, if you're still reading, you're curious AND smart,
So take one more leap back to the start —
In the 4th century BC,
Thrived a special society
Where knowledge and art were on the increase.
The name of this place was ancient Greece.

Here lived ARISTOTLE, a man of learning,

Stroking his beard, ideas churning.

He was a teacher to the king,

He tried to study everything,

Pointing to the stars as they were twinkling,

He said, "All can be sorted just by thinking.

Who needs experiment

When you have judgment?

As sure as trees are firewood, and sure as cows are leather,

Man rules all, because he's so clever."

He gathered his pupils on nights so clear
To point out that we live on a special sphere.
As the sun went down and the night was chill,
He believed the earth was absolutely still.
He said, "The heavens are perfect and never changing
For the stars don't go about rearranging.
I'm sorting out all this jumble,
So we can look at the sky without feeling humble.
For though we live in a universe small
Of course, we are the center of it all."

Yet only one hundred years on,
When Aristotle himself was gone,
ARISTARCHUS argued, "The sun is huge
And around this sun our small planet moves."

But the late great Aristotle was so respected,
Aristarchus's findings were not accepted.
The people cried, "We prefer good old Aristotle."
And so the ideas of Aristarchus were throttled.

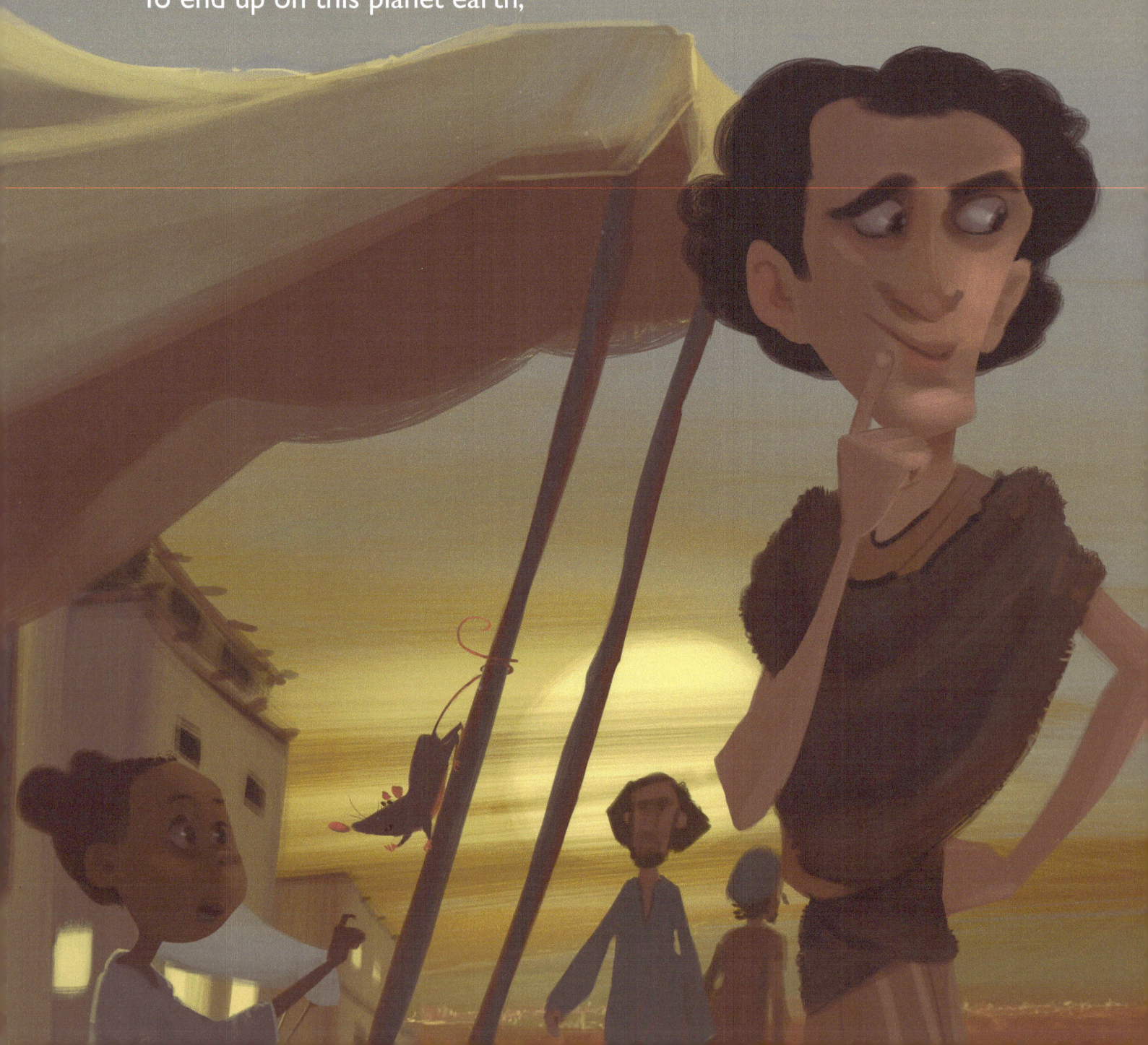

Oh curious child,

If you've ever looked at the sky in wonder, then follow me,

To another ancient world where now we meet PTOLEMY,

Rome was in charge, ninety was the year, and Alexandria, Egypt the place.

Ptolemy drew a great map of the world, but messed up outer space.

With Aristotle, Ptolemy agreed.

In his writing he did concede,

"You are lucky children, by your birth,

To end up on this planet earth,

The universe is not a riddle,

When you are sitting in the middle.

There are no more questions, nothing to disprove

The sun revolves around us, our planet does not move."

Though Ptolemy was very wise

The truths he claimed turned out to be lies.

Then came along, one thousand four hundred years later,

A quiet Polish man, just that bit greater.

Now I don't like to moan or whine,

But NICOLAUS COPERNICUS, you took your time!

Curiously, he gazed at the sky so long,

Observing the planets, but it all seemed wrong.

"Old Aristotle and Ptolemy must have been a little dense
For when I add it up it makes no sense.
Unless these stars are doing acrobatics
They don't fit in with mathematics."
He wrote down his findings in a book
And let a few friends take a look.
Soon there were whispers, "The truth will advance;
Our spinning home joins the other planets in a dance.
What you see is not what you get,
The sun does not rise, nor does it set.
We revolve around our sun, the great force of light.
Nothing is still, life is in flight.
Every 365 days we journey around this sun in a ring
That gives us summer, fall, winter and spring.

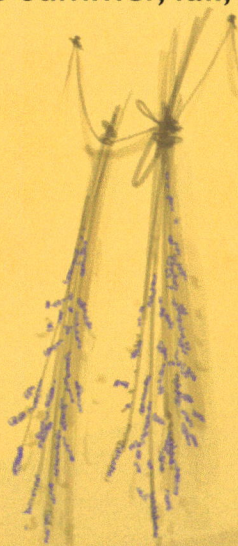

We are spinning, we are turning,

In the center, the bright sun is burning.

To one great life force, all planets are in thrall,

—Yes, the burning, bright sun IS the center of all."

The powers of the time didn't take this news lightly
To be at the center made them feel mighty,

They huffed
They puffed
They chanted
And ranted,

"Nicolaus Copernicus,
A man most ridiculous
Ridiculous Nicolaus
Copernicus ridiculous

It's really most fractious
To move on your axis.
Hold on tight, you could float away,
If we all keep turning once a day.
Don't tell the teachers, they'll tell the young,
He's moved the earth and stopped the sun."
Nicolaus Copernicus did not want a riot,
Frightened of the powers, he kept it all quiet,
This gentle man wasn't keen on fighting
So he hid away and kept on writing.
Thankfully, his friends took his work,
And printed a most magnificent book.
Of the great man they were devoted fans,
They carried the book back and placed it in his hands.
Oh curious child, it is what you fear
The lies lived on for many a year,
Though his book survived, poor Copernicus was daunted
That's the way with truth: Often it's not wanted.

GALILEO GALILEI — an Italian boy with hair of flaming red,

Mastered music, painting, math, so many books he read,

By the time he was a young man there was much about the world he knew,

And when he grew a bushy beard, why that was red too!

He said: "What we know of the world will be but a fragment

If we don't learn to repeat every experiment,

All our work will be deniable

Unless our findings are reliable

For once a truth is found, it would be a shame

If we can never find that truth again."

After watching pendulums swing and measuring rolling balls

Galileo felt the great sky call.

But the trouble with each and every star

Is that it sparkles in a sky so far.

To study them all he didn't have a hope

So Galileo designed his own telescope!

At home in Florence he could now see clearly
Things that would wind up costing him dearly.
In his garden, as the day grew dimmer,
He delighted to see the Milky Way shimmer.
He saw that the moon had mountains, and the sun had spots,
But what he saw next would cost him a lot,
For around the planet Jupiter he saw four moons move
And if they circled another planet, that went on to prove
Not everything around our planet earth revolves.
So the great mystery began to be solved.
For though Copernicus's ideas were not allowed,
Galileo Galilei declared clear and loud,

"Nicolaus Copernicus,
A man not so ridiculous
Not so ridiculous this Nicolaus
Copernicus not ridiculous
It's not at all fractious
To move on your axis.
We are spinning, we are turning,
In the center the bright sun is burning,
To one great life force all the planets are in thrall.
Yes, the burning bright sun IS the center of all."

In those days you couldn't just upload a blog

Instead he wrote a book — The Dialogue.

The church horrified by his position,

Brought Galileo before the Inquisition:

They came knocking on the door of his home

And dragged the great scientist all the way to Rome.

Galileo was old and sick,

This was not a fair fight to pick.

Galileo, his red beard now turned grey,

Was worried he'd never see the light of day

Though he knew the truth, he went very pale

At the thought of life in a Roman jail.

He shrugged, "I am here in your hands, do with me what you please."

So they dressed him in white and put him on his knees.

They chanted, "Galileo Galilei — go away go away

Galileo Galilei — be silent and obey

Galileo Galilei — you will pay, you will pay."

Frightened of these men, he cried, "You are right!
I will no longer put up a fight."
They had the sick old man ensnared
And passing a sentence, they declared,
"In sixteen hundred and thirty three,
Galileo Galilei you will never again be free.
Galileo Galilei — you must be quiet as a mouse
Galileo Galilei — you will be imprisoned in your house."

The Inquisition were a bunch of rotten stinkers,

But in this world there will always be free thinkers.

Galileo kept working, though he lost his sight,

He knew what the truth was, he knew he was right.

For even though you might go blind,

You can still see clearly in your mind.

And though his book was banned and driven underground,

Underground is where seeds are found

And soon those seeds would be sprouted,

Others would recognize truth and shout it.

Working near Prague was a Dane named **TYCHO**

Who had a fight with his cousin that went a bit psycho.

What caused such a fight? What provoked such wrath?

It wasn't a person, but a problem in math.

The cousins argued for weeks and weeks,

You have to remember they were a couple of geeks.

They met at midnight to fight a duel,

Poor Tycho ended up the fool.

The unlucky man fell back and roared

When his cousin sliced off his nose with a sword.

Life is hard you might suppose

When you are left without a nose.

Soon Tycho was back in fine fettle,

Full of ideas, sporting a nose of metal.

When he became astronomer to the king,

Astrology and astronomy were thought the same thing.

His advisors said: "The future is uncertain so why wait?

When the stars hold the secrets to our fate.

You may have a brass nose, poor Tycho, but knowledge is your gem,

For all kings believe the heavens are actually about them.

While astronomy is a science, astrology is an art,

Who better to combine them than someone so smart."

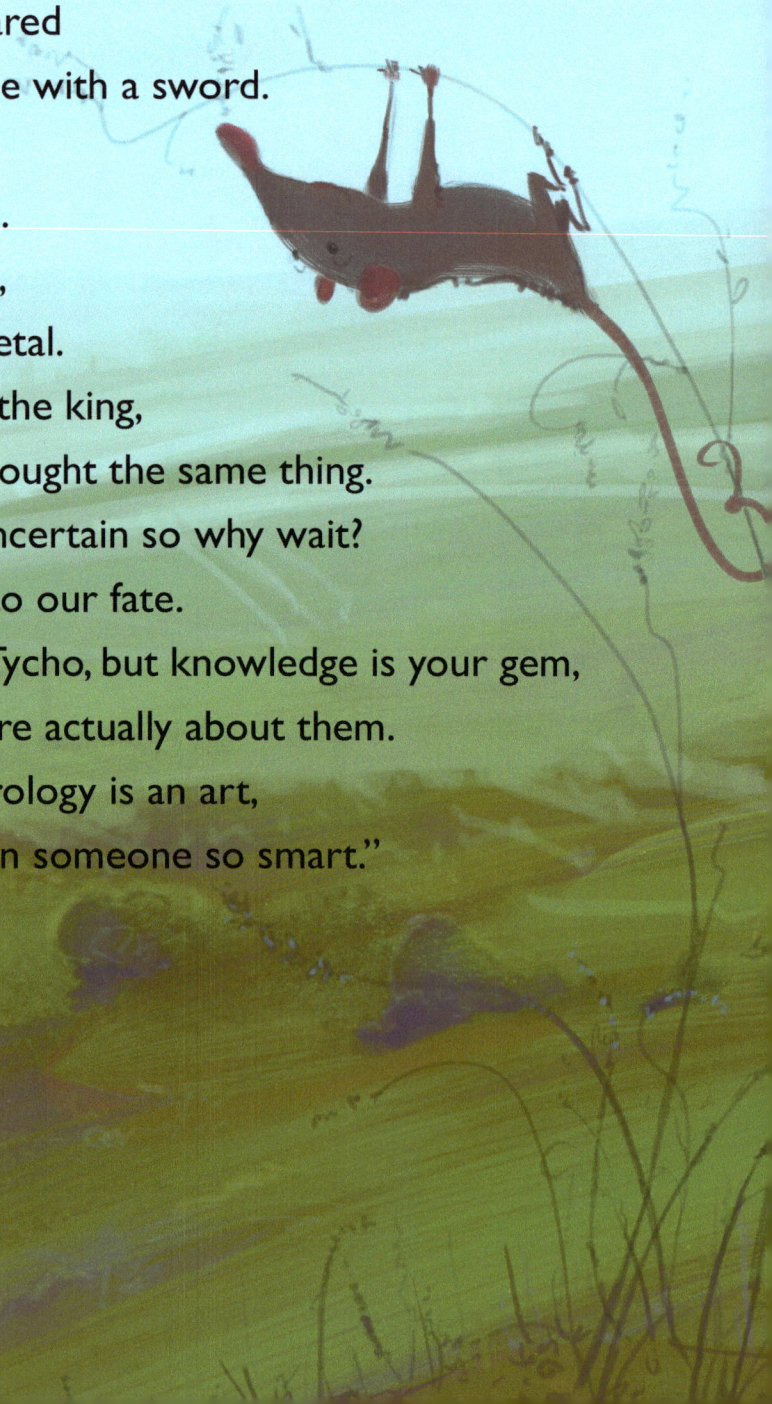

He swaggered around with great freedom
Fancying himself the cleverest in the kingdom,
The king lavished him in honors and hard cash,
But in Prague he would soon meet his match,
With his fake nose, he struck all as a curious creature,
And soon there would be a student smarter than the teacher.

Little JOHANNES KEPLER was often ill
But had a mind that could not be still.
He came from Germany and it was a surprise
That Kepler was the one to sort truth from lies.
His father left for war, never again to be seen,
So his grandparents brought him to live in their Inn.
His prospects were as dim as his nose was runny,
Never in his life would he have any money,
His skin was scarred with the dreaded pox,
His hands so weak he couldn't put on socks.
To top it all, his sight was bad—
He had plenty of reasons to feel sad.
But when in the Inn he chatted to guests,
They couldn't help but be impressed,
For no sooner had they walked in from the path,
Then the little boy would dazzle them with math.
As they put their elbows on the table
They saw he was bright and able.
Yet even these admiring travellers could not foresee
How extraordinary his life was to be.

In sixteen hundred and four,
Looking up at the night sky he let out a roar.
For there appeared to his delight
A brand new star, so sure and bright.

Remembering a comet when he was a kid
He told all, "We do not live in a jar on which there is a lid.
Out there there's something bigger and ever changing.
The heavens aren't still, but always rearranging."

Eagerly, Kepler packed a bag
And left to work for Tycho in Prague,
But Tycho's nose was as cold as his heart
When he saw young Kepler was beyond smart.
Tycho distracted Kepler with problems so hard,
And his own calculations, he kept under close guard.
He ordered Kepler to concentrate on Mars.
While he himself took on the rest of the stars.
But as his boss worked for glory.
Kepler was about to change the story:
In his close studies of Mars's movement,
The teacher was eclipsed by the student.
Kepler's eyes were weak but his vision was bigger.
He figured out something none of the rest could figure,
Clapping his hands and licking his lips,
He proclaimed: "Planets don't move in circles, but in an ellipse."
And although at first it seemed farfetched—
An ellipse is only a circle that's been stretched—

Kepler's body wasn't strong, but his mind was sky bound.
The evidence for Copernicus's ideas he found,
Though Galileo could only guess by observations
All was finally proven by Kepler's calculations.
For once and for all
The big and the small
The powerful and the meek
The strong and the weak
Had to shrug their shoulders and admit
Kepler's math made it all fit.
Around the sun all the planets move,
Johannes Kepler finally did prove
We are spinning, we are turning,
In the center the bright sun is burning,
To one great life force all the planets are in thrall,
Yes, that burning bright sun IS the center of all.

You might wonder why, oh curious one
Why such a fuss over stars, moon and sun?
Why, oh why, do grown-ups cling to ideas that are wrong?
Because they want to feel powerful, they need to feel strong
No one likes to feel so little,
Everyone would rather be right in the middle,
That's why the powerful became all whiny
When told they lived on a planet so tiny.
Though the stars are much more than a beautiful jumble,
When we look at them, it's ok to feel humble.
Even if the magical night sky has been measured
This small planet earth is our home, and must be treasured.

You have just read a small history
Of how we unravelled a big mystery.
In this book it was shown
No one person discovered truth alone,
It took several clever people over a very long time
To come up with answers so sublime.
It happened not in one time, nor in just one nation—
Discovery takes hard work and co-operation.

There's so much more to know, lots of ideas to be grasping
My clever child let your curiosity flow — never EVER stop asking.

SUN

MERCURY

VENUS

EARTH

MARS

JUPITER

SATURN

URANUS

NEPTUNE

EARLY ASTRONOMERS

Aristotle
(384–322 BC)

A Greek philosopher and scientist, Aristotle was one of the greatest thinkers in all of ancient civilization, whose studies laid the foundation for scientific thought. His writings covered many different subjects: poetry, music, logic, politics, biology, zoology, physics, and metaphysics. Together with Plato, who was his teacher, and Socrates, who was Plato's teacher, Aristotle had a vast influence on the philosophy of Western civilization. At the age of 17, he was sent from Stagira, where he was born and raised, to Athens, to study philosophy at Plato's Academy. He stayed there for the next 20 years. Although Aristotle was an exceptional student, he opposed some of Plato's teachings. In 342BC, Aristotle was invited to tutor Alexander the Great. After he returned to Athens he started a school named Lyceum. His method became known as *peripatetic* (Greek: walking about), because Aristotle taught his students while walking. All of Aristotle's ideas fitted with the belief of his time that earth is the center of the universe and supported geocentric theory (Greek: *gea*-earth). In Aristotle's time, ordinary people believed that the earth was flat. Aristotle made a big contribution to astronomy in explaining why the earth had to be spherical. Aristotle noted that the earth was casting a shadow on the moon during the lunar eclipse and that shadow was always the same arch shape, like the edge of a circle. He also observed that if a person was traveling to the north or the south from his present location, the positions of stars were shifting, with some constellations disappearing below the horizon. That can happen only if one is moving on the surface of a sphere. Today we still refer to Aristotle's classifications that define and describe basic scientific principles (matter, energy, species...).

Aristarchus
(~310–230 BC)

A Greek astronomer and mathematician born in Samos, Aristarchus was the first person to present the heliocentric model with the sun at the center and the earth revolving around it. Aristarchus's ideas and heliocentric (Greek: *helios*-sun) theories were often rejected in favor of the geocentric theories proposed by Aristotle and later on by Ptolemy. Because his work was destroyed, little is known about Aristarchus from his originals. However, we know a bit from books by other authors, e.g., Archimedes (287-212BC). Aristarchus was a student in the Lyceum school established by Aristotle. In his studies, Aristarchus resolved how to measure relative distances and sizes of the sun and the moon. Aristarchus's only surviving work *On the Sizes and Distances of the Sun and Moon* provides geometric arguments and calculations of the moon's, sun's and earth's diameter. Aristarchus calculated that the sun is about 20 times farther from the earth than the moon, and around 20 times bigger than the moon. In fact, the sun is around 400 times bigger than the moon, and around 400 times farther away; the two just appear to be the same size when observed from the earth. It is important to state that the errors in Aristarchus's measurements were not due to his lack of knowledge or understanding of the universe, but, to the limitations of available instruments. His successors, Hipparchus and Ptolemy, refined Aristarchus's methods and calculated very accurate values for the size and distance of the moon. However, all ancient research underestimated the size and distance of the sun and did not agree with Aristarchus's heliocentric theory; thus, the geocentric theory prevailed for next 1700 years. Aristarchus's heliocentric ideas were finally revived by Nicolaus Copernicus.

Ptolemy
(~87–170 AD)

A Greek-Egyptian mathematician, astronomer, and geographer born in Alexandria, Ptolemy proposed the first general theory of cosmology. Ptolemy systemized the Greek geocentric view of the universe, and rationalized the apparent motions of the planets, as they were known at the time, using very accurate calculations. Thanks to Ptolemy's findings, medieval astronomers were remarkably good at astronomical predictions. The majority of Ptolemy's theories of the universe are described in his book *Mathematical Syntaxis*, also called *Almagest*. *Almagest* contains information about the earth, sun, moon, and stars' movements, as well as eclipses, and a breakdown of the length of months. It also included a star catalog containing 48 constellations, with the names we still use today. Ptolemy changed the model of the planetary movement established by earlier Greek astronomers. His new "Ptolemaic model" explained motions of the planets by assuming that each planet moved in a small sphere or circle (an epicycle), which itself orbited along a larger sphere (a deferent), with earth in the center. Ptolemy believed that the planets and the sun orbited the earth in this order: Mercury, Venus, Sun, Mars, Jupiter, and Saturn. The Ptolemaic system is based on the work of Greek and Babylonian astronomers—mainly Hipparchus (~190-120BC). In addition to being a great astronomer, Ptolemy made a significant contribution to geography, proposing the use of longitude and latitude in order to determine the points of the globe. Ptolemy's maps covered about a quarter of the earth's surface: from the Canary Islands on the east, to China on the west, from the Arctic on the north, to Africa on the south. Even the Italian explorer Christopher Columbus (1451-1506) used Ptolemy's maps on his voyages.

Nicolaus Copernicus
(1473–1543)

Copernicus was a Polish renaissance mathematician and astronomer who proposed a heliocentric model of the solar system, in which the planets orbit around the sun. Even today, this idea is known as the 'Copernican Theory'. Copernicus studied first to be a priest, then he studied law and medicine, but his true passion was astronomy. In 1513, he built his own observatory. Soon after, he distributed his handwritten book *Commentariolus* (Latin: *Small Commentary*) in which, for the first time in history, using very accurate mathematical formulas and calculations, he placed the sun at the center of the universe. He suggested that the earth's rotation was what accounted for the rise and setting of the sun, the movements of the stars, and the seasons. During his research, Copernicus came to the conclusion that stars were much farther away than the sun, and that it was impossible for stars to move around the earth every 24 hours, because in that case, the distant stars would need to travel at lightening speeds around the earth. He proposed the circular path of the planets around the sun. He also proposed that the earth rotates around its axis. He stated that the earth revolves around the sun once a year, and that the earth is the center of the moon's orbit. Copernicus was the first to explain that the universe was big and changing rather than fixed. It wasn't until 1543, near the end of his life, that Copernicus published his findings. It is believed that Copernicus delayed publication of his book due to fear of persecution from the Church and the Pope. His findings that put the sun in the center of the universe did not support the dogmas of the Catholic Church. Copernicus's ideas would, in the future, lead to a better understanding of the universe and gravity, starting the 'Copernican Revolution' which was followed by the 'Scientific Revolution'. Copernicus is thus considered the founder of modern astronomy.

Galileo Galilei

(1564–1642)

An Italian astronomer, physicist, and mathematician, Galileo made profound scientific contributions in those fields, and played a major role in the beginning of the movement known as the 'Scientific Revolution'. While a young man, Galileo considered becoming a priest. In 1581, he enrolled in the University of Pisa to obtain a medical degree, but at that time medicine was stuck in the old Greek methods of Aristotle and Hippocrates. He was more interested in the mechanics of a swinging pendulum, so he became a professor of mathematics at the same university. In 1592, he moved to the University of Padua as a teacher of mathematics, geometry, and astronomy. His biggest interest was the universe and he was looking for the ways to study it. Though primitive versions of the telescope were available at the time, he built his own telescope with magnification power of 30x, and was the first person to apply the telescope to astronomy. Using his telescope, Galileo observed craters and mountains on the moon, sunspots, four moons orbiting around Jupiter, and the phases of Venus. The Catholic Church lauded Galileo's discoveries of Jupiter's moons. However, when he suggested that the earth orbits the sun in the same fashion as Jupiter's moons orbit Jupiter, the Church accused him of heresy. At that time there was an organization within the Catholic Church called the Inquisition. The powerful Inquisition's main job was to make sure no one contradicted the Church's teachings. At the same time, Galileo wrote an extremely accessible book for a wider audience called *The Dialogue*, explaining the science behind heliocentric theory. In 1633, he was sentenced by the Inquisition to house arrest until the end of his life. He was scared, old, and sick, so he retracted his views and declared publicly that the fixed earth was at the center of the universe. Legend has it that he whispered, "But yet it moves."

Tycho Brahe

(1546–1601)

Tycho was a Danish astronomer and alchemist. At the age of twenty, during his studies at Germany's University of Rostock, Tycho lost his nose in a sword duel with a fellow student who was also his cousin. The two were fighting over a mathematical formula. After this event, Tycho had to wear a replacement nose made of copper and brass. The amazing sight of a predicted solar eclipse in 1560 changed Tycho's life and turned his interest to the skies. He built an observatory, and in 1572, reported a supernova. Shortly after he became famous for disproving the Aristotelian theory of a never-changing universe. For this achievement, the Danish King, Frederick II, gave him a small island, called Hven, and provision for his work. On the island, Tycho built two observatories equipped with improved and enlarged instruments for studying celestial movements. In the 1570s, Tycho postulated his own 'geo-heliocentric' system in which the sun and the moon orbited the earth, while the other five known planets orbited the sun. For Tycho's time, that was a very safe proposal. Tycho's theory did not violate the laws of physics, or the sacred scripture (in contrast to Copernicus's geocentric theory that was, at the time, unaccepted by the very influential Catholic church). Tycho moved to Prague where, in 1600, he met Johannes Kepler. Just one year later, he died of a sudden infection. There are speculations about his death due to high amounts of mercury found at the tips of his hair. Was it his brass nose, or was he poisoned? We will never know. What we do know, however, is that he jealously kept his calculations from Kepler. Kepler benefited from his death, as he finally was able to study them in their entirety.

Johannes Kepler
(1571–1630)

Kepler was a German mathematician and astronomer. As a child, he was sickly and small due to his premature birth and frequent illnesses. At the age of three he was almost killed by Smallpox. His tiny, frail appearance belied his intelligence and strength of mind. In 1587, Kepler was offered a scholarship at the University of Tubingen to study ministry, philosophy, mathematics, and astronomy. During his studies, young Kepler was astounded by Copernican theories. In 1596, while teaching mathematics in Graz, he wrote the first outspoken defense of Copernican theory in *Mysterium Cosmographicum*. In this book he supported Copernicus's ideas of the heliocentric solar system and he modernized them with new models and calculations of planetary movements. In 1600, Kepler moved to Prague to become an assistant to Tycho Brahe. At that time, royalty employed astronomers as astrologers, and used their knowledge of the stars to predict events.

Tycho, perhaps in fear of being over-shadowed by his brilliant assistant, was reluctant to share his findings with Kepler. Therefore, as his teacher, he instructed Kepler to work only on the orbit of Mars. In 1601, Tycho died suddenly and Kepler inherited his meticulously written observations, with over a thousand carefully mapped celestial bodies. Kepler used Tycho's data to determine the nature of planetary orbits. Being an excellent mathematician and astronomer, Kepler realized that planets are moving around the sun in elliptical rather than circular orbits. Therefore, he proposed three scientific laws that descibed the motions of planets around the sun, today known as 'Kepler's Laws of Planetary Motion'. Kepler's astronomical accomplishments laid the foundation for Isaac Newton's (1642-1727) theory of gravity. Along with Kepler's deep understanding of astronomy, he was fascinated with optics. Kepler invented his own enhanced version of the refracting telescope. He also devised eyeglasses for farsightedness and nearsightedness. Today, besides being one of the major astronomers, Kepler is also known as the founder of modern optics.

DICTIONARY

AD: Anno Domini, Latin for Year of the Lord.

Astronomy: The study of celestial objects such as galaxies, stars, moons, and planets.

Astrology: The study of the assumed influence of planets and stars on human affairs.

Axis: A straight, imaginary line about which a body or a geometric figure rotates.

BC: Before Christ.

Celestial: Most commonly relating to the sky or the heavens. Planets are celestial bodies.

Comet: A frozen mass of dust and gas revolving around the sun leaving a tail-like structure.

Constellation: A grouping of stars named by ancient astronomers after shapes they resemble.

Cosmos: All existing matter and space; also known as the universe.

Cosmology: The study of the universe.

Earth: The third planet from the sun in the solar system. We live here!

Ellipse: A shape that resembles a flattened circle (an oval).

Galaxy: A group of billions of stars held together by gravity.

Geocentric theory: An idea that places the earth at the center of the solar system.

Gravity: The force between objects that have mass.

Heliocentric theory: An idea that places the sun at the center of the solar system.

The Inquisition: A powerful and feared organization within the Catholic Church during the Middle Ages.

Jupiter: The fifth planet from the sun, and the largest planet in the solar system.

Mars: The fourth planet from the sun.

Milky Way: The galaxy which contains the solar system in which we live.

Moon: A natural satellite orbiting a planet.

Orbit: The path of an object around its sun or planet.

Pendulum: A weight hung from a fixed point so that it can swing freely backward and forward.

Planet: An object moving around its star.

Revolving: When one object is moving in a circle or ellipse around another object. The moon revolves around the earth; the earth revolves around the sun.

Rotating: The spinning of a spherical object. The earth rotates around its axis.

Satellite: A small object orbiting a larger one. The moon is the earth's satellite.

Solar System: A system of planets and other objects orbiting the sun.

Sphere: A round, solid object.

Star: An object in space made of burning gas that can be seen in the night sky.

Sun: The central body of the solar system.

Sunspots: Regions of lower surface temperature on the sun that form dark patches.

Telescope: An optical instrument designed to make distant objects appear nearer.

Universe: All existing matter and space; also called cosmos.

Emer Martin is a Dubliner. Her first novel *Breakfast in Babylon* won Book of the Year 1996 in her native Ireland. Houghton Mifflin released *Breakfast in Babylon* in the U.S. in 1997. *More Bread Or I'll Appear*, her second novel was published internationally in 1999. Emer studied painting in New York and has had two sell-out solo shows of her paintings at the Origin Gallery in Dublin. Her book, *Baby Zero*, was published March 2007 and internationally in 2014. Emer was awarded the Guggenheim Fellowship in 2000. She has two young daughters and lives between the jungles of Co. Meath, Ireland and the valleys of Silicon in California. She founded the independent publishing cooperative Rawmeash. This is her first children's book. You can follow her on emermartin.tumblr.com or on twitter@emermartin. Her website is www.emermartin.com

Suzana Tulac was born and raised in Croatia. She moved to California in 1997. Her Ph.D. in molecular biology was a result of collaboration between Stanford University, where she worked for five years, and the University of Zagreb, Croatia. As her three kids entered elementary school she became actively involved in children's science education. Over the last few years she attended many talks and workshops on the history of science. She realized there is a lack of fun, accessible materials on this subject for children which prompted the idea for this book. In addition to science, she is very passionate about photography. She lives with her family in Mountain View, CA. You can visit her at www.suzanatulac.weebly.com.

Magdalena Zuljevic is also known as Magi. Born and raised in Croatia, Magdalena studied at The Academy of Fine Arts in Zagreb, earning a BFA in Art education. After experimenting with sculpting and oil painting, she decided that illustration was her true calling. Now Magi specializes in illustrating for the children's market. She currently lives with her husband, two children, her parrot Cody, and her guinea pig Sparkle in Sunnyvale, CA. See what she is up to at www.pencilfairy.com.

www.ingramcontent.com/pod-product-compliance
Lightning Source LLC
Chambersburg PA
CBHW041220040426
42443CB00002B/27